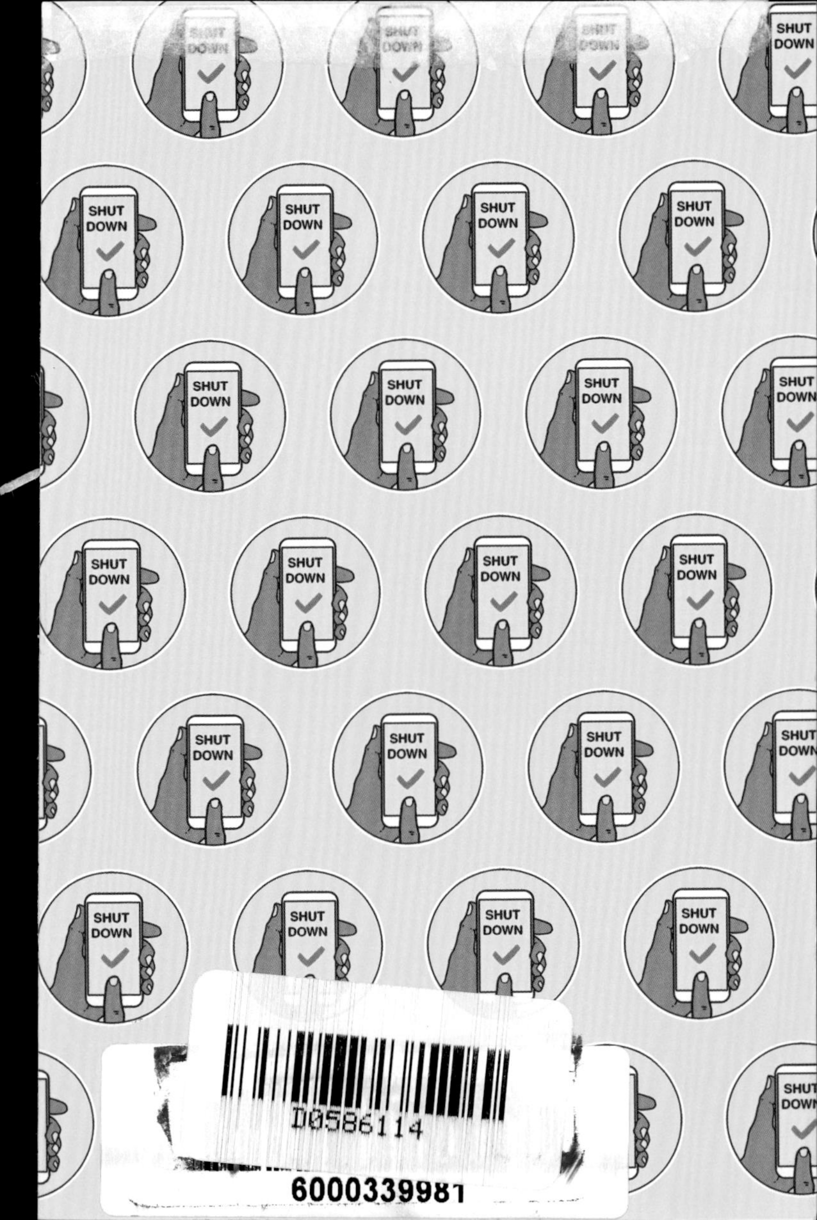
SHUT
DOWN

STOP LOOKING AT YOUR PHONE

STOP LOOKING AT YOUR PHONE
A HELPFUL GUIDE

ILLUSTRATED BY SON OF ALAN

Pop Press

1 3 5 7 9 10 8 6 4 2

Pop Press, an imprint of Ebury Publishing
20 Vauxhall Bridge Road
London SW1V 2SA

Pop Press is part of the Penguin Random House group
of companies whose addresses can be found at
global.penguinrandomhouse.com

First published in the United Kingdom by Pop Press in 2018

www.penguin.co.uk

A CIP catalogue record for this book is available from the
British Library

ISBN 9781785039096

Printed and bound in Great Britain by Clays Ltd, St Ives PLC

Introduction

Do you ever idly click open your phone without actually knowing what you want from it? Do you ever feel anxious when you realise you've left it charging at home? Have you ever had text neck (now a genuine medical condition) or felt your thumbs seize up from swiping left? Do you sometimes look up from scrolling through Instagram sunsets (#Blessed #Sunset *prayer hands emoji*) to realise that it's half past two in the morning? If so, *Stop Looking at Your Phone* is the perfect guide to help you re-engage with the world around you.

We all know phones are useful. They get us home late at night, keep us in touch with friends and up to date with current events all over the world, and help us to find out instantly what happened to that one child star from the 1990s who came up in the pub quiz the other day. Our parents could never have dreamt of a world where anyone could own a portal to an endless supply of kitten videos.

But, recently, maybe you've had the sense that your phone has become a little ... demanding. Perhaps it's buzzing away right now, insisting that you check out the photos of all your friends having brunch (#avocado #nomnomnom) while you've just opened the fridge to find half a sad looking carrot. And could it be that you occasionally get the feeling that you are missing out on things? Like the time at that music festival, when you spent so long working out how to film the gig that afterwards you couldn't actually remember any of the songs.

Stop Looking at Your Phone is a compendium of practical hints and tips to help you put your phone or other handheld electronic device in your pocket and re-establish contact with the people around you. From how to celebrate birthdays to what to do if you find an injured dolphin, this handy pocket-sized guide walks you through real life situations, using clear illustrations for every occasion. It could save you from a grizzly and embarrassing death or, at the very least, marginally improve your life.

CONVERSATION

A.
DO U WNT
....
....

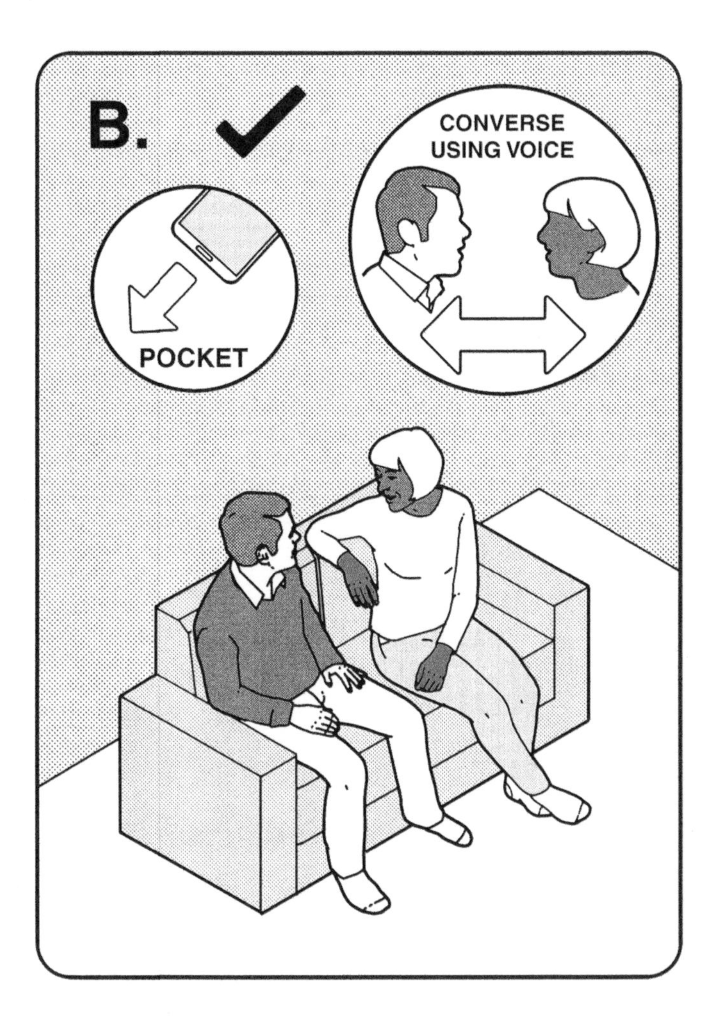
B.
CONVERSE
USING VOICE
POCKET

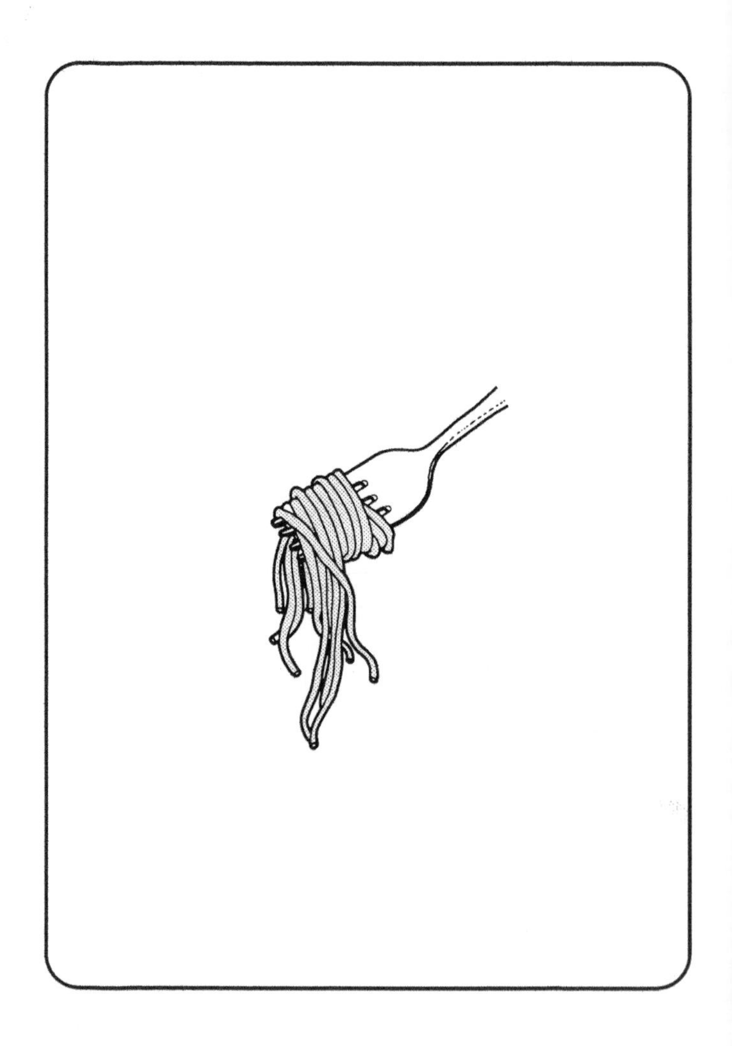

EATING OUT

A.
INSTAGRAM
PICTURE
FILTER
EFFECTS

B.
1. EAT FOOD
BEFORE IT GETS COLD
2. CHEW PROPERLY
POCKET

BURGLARY

A.
WARCRAFT
SLAYER
LEVEL:
6890

B.
DISENGAGE
FROM PHONE
ENJOY
LOWER CHANCE
OF BURGLARY

THE BABY

A.
PHOTO
UPLOAD
INSTA-SNAP
INSTA-SNAP

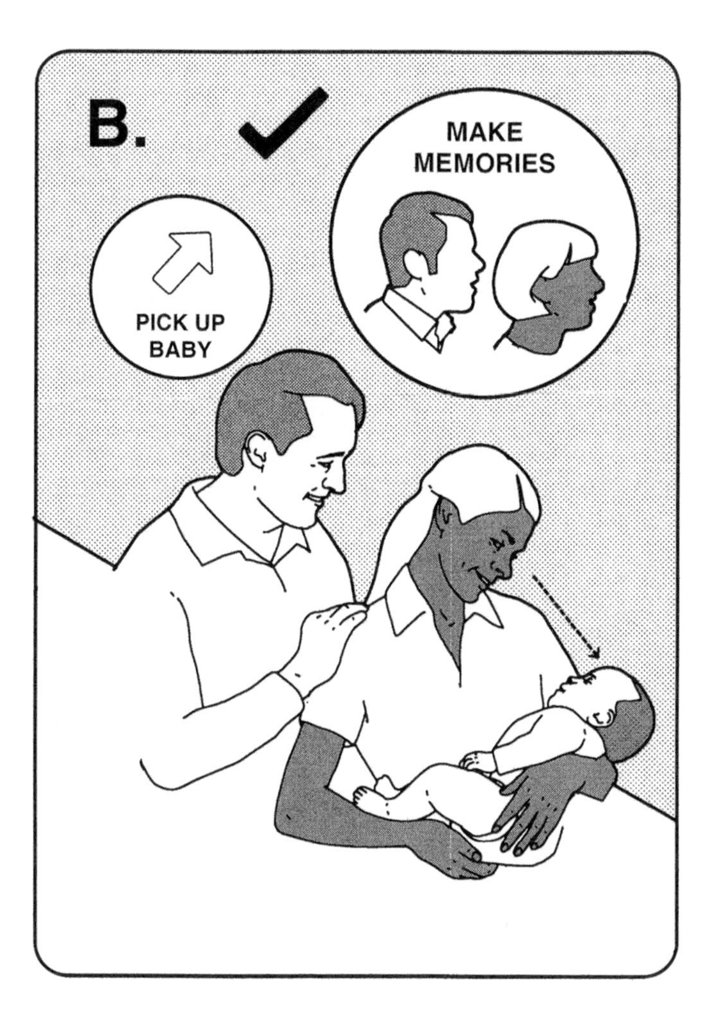
B.
MAKE MEMORIES
PICK UP BABY

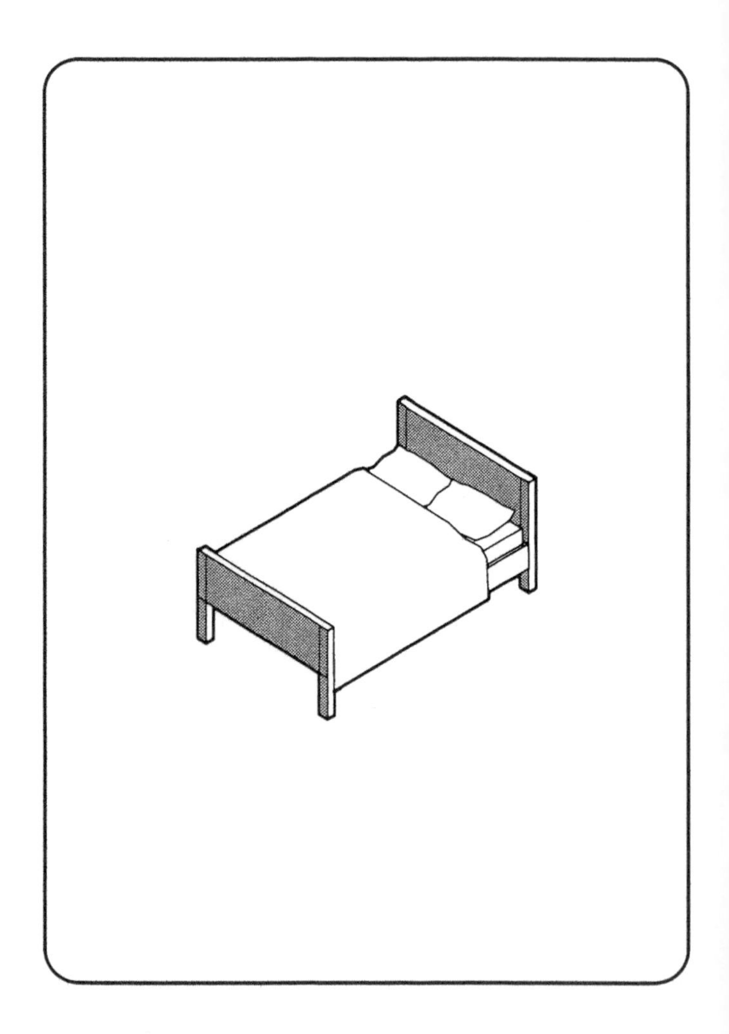

THE BEDROOM

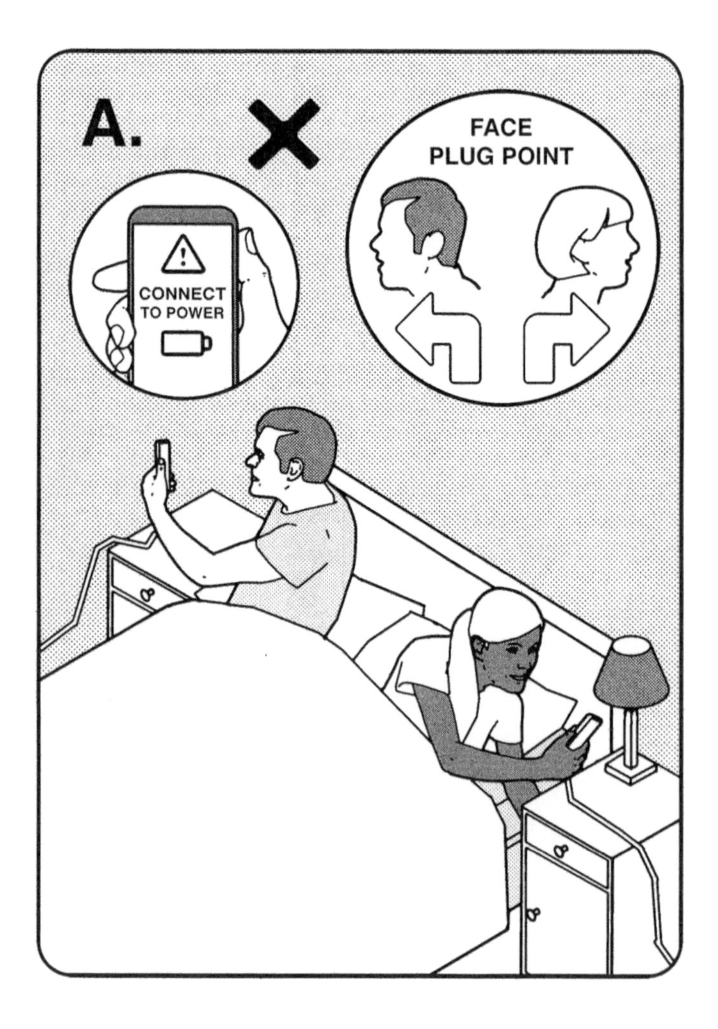
A.
CONNECT
TO POWER
FACE
PLUG POINT

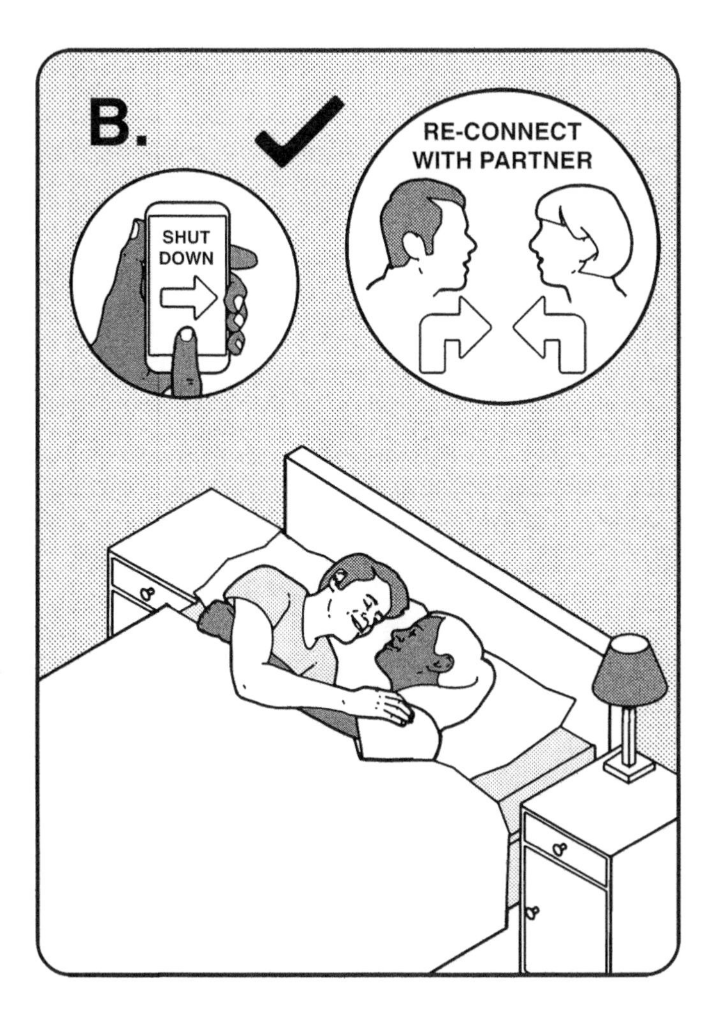
B.
SHUT
DOWN
RE-CONNECT
WITH PARTNER

FAMILY TIME

A.

B.
INTERACT
WITH FAMILY
SHUT
DOWN

FINDING LOVE

A.
LOVE
MATCH
SOUL
MATE
FINDER

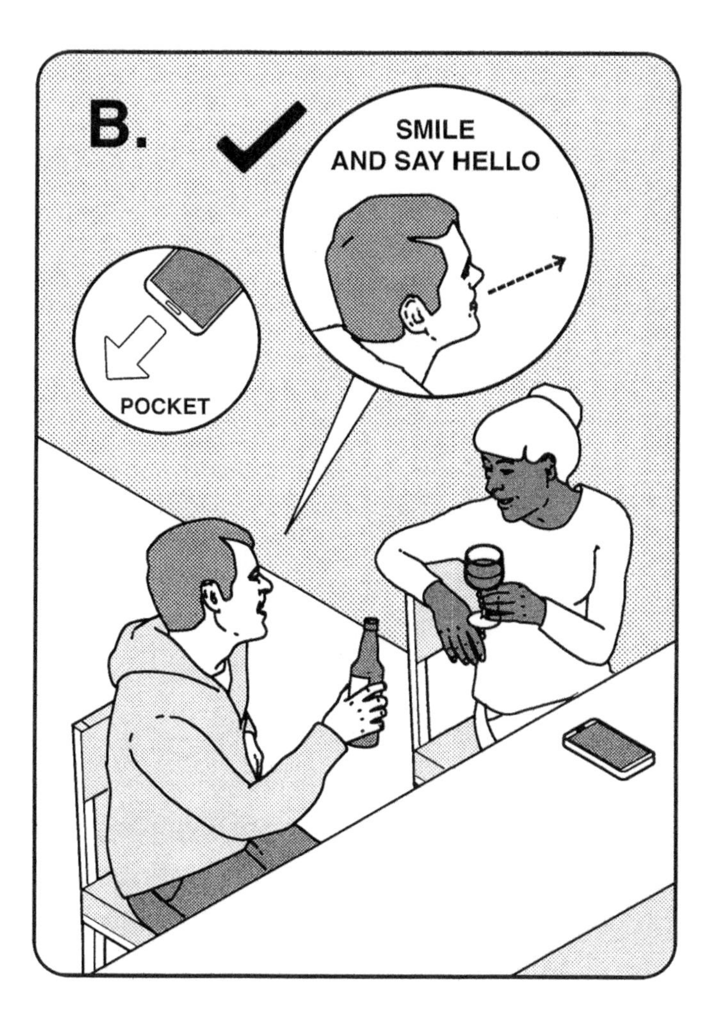
B.
SMILE
AND SAY HELLO
POCKET

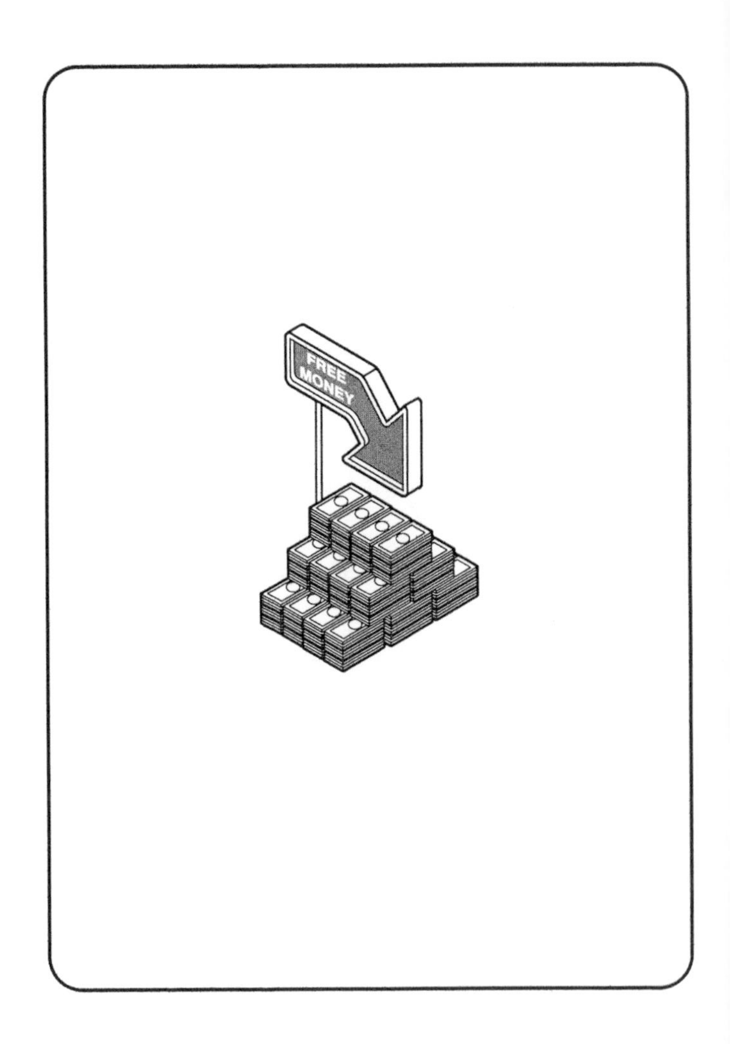
FREE
MONEY

FREE STUFF

A.
FREE MONEY

B.
LOOK UP!
FREE MONEY

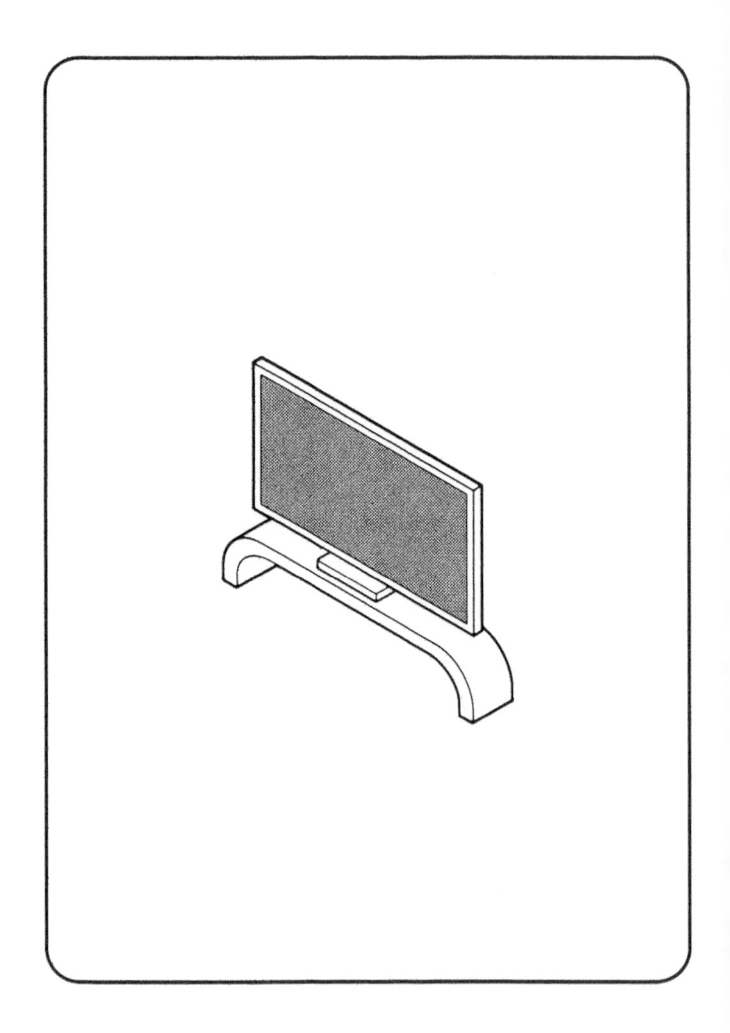

THE GOAL

A.
FOOD SHOPPING LIST
FANTASY FOOTBALL TEAM
GOAL!!

B.
SHARE AN
EXPERIENCE

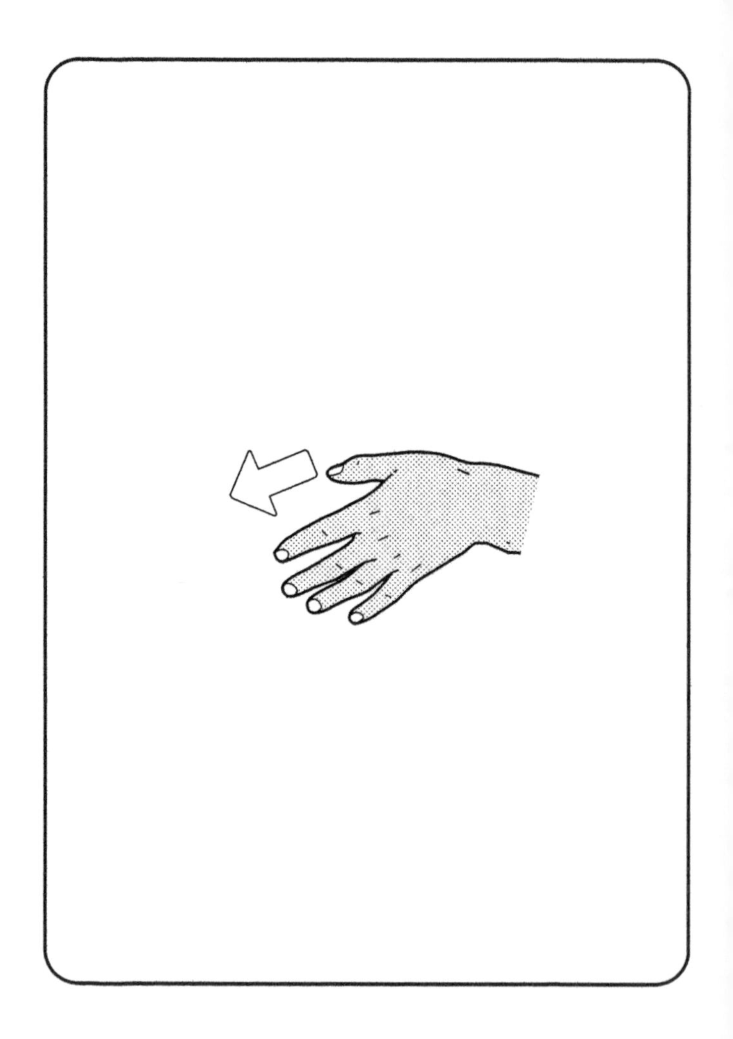

HELPING HAND

A.
SELFIE
OPPORTUNITY

B.
ENJOY
FEELING
OF SELF WORTH
HELP
DYING CREATURE

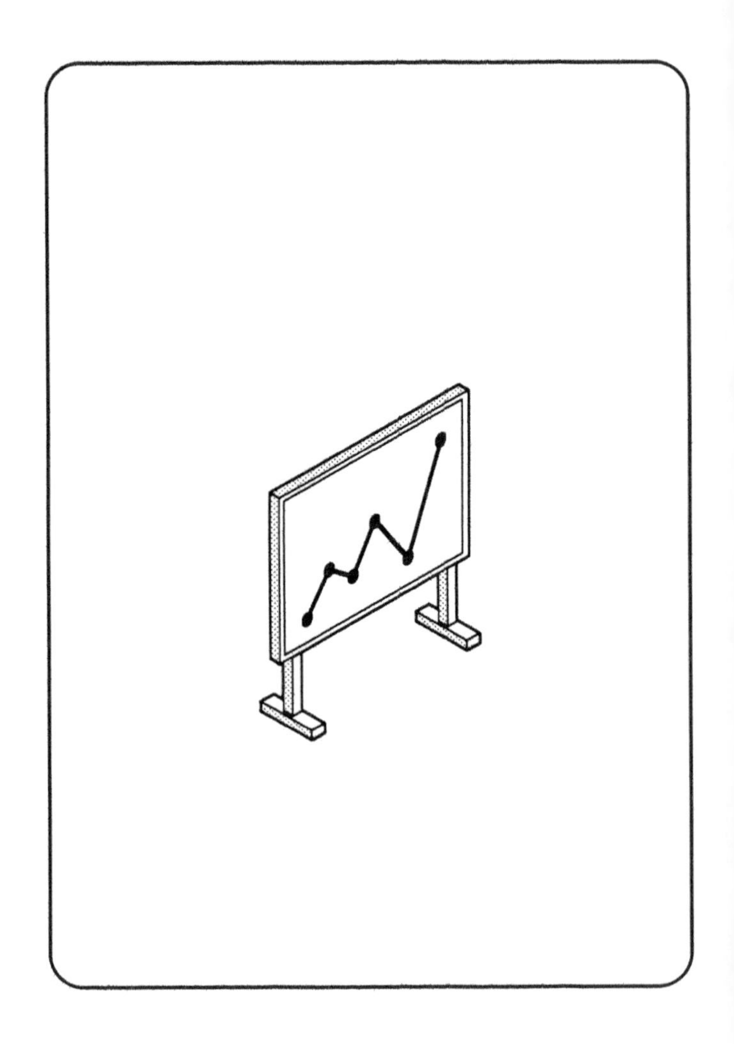

THE MEETING

A.

B.
ENJOY
ADMIRATION
OF COLLEAGUES
POCKET
X+Z = GOAL
X
Z

THE NIGHT OUT

A.
SELFIE
ACCOUNT
FULL!

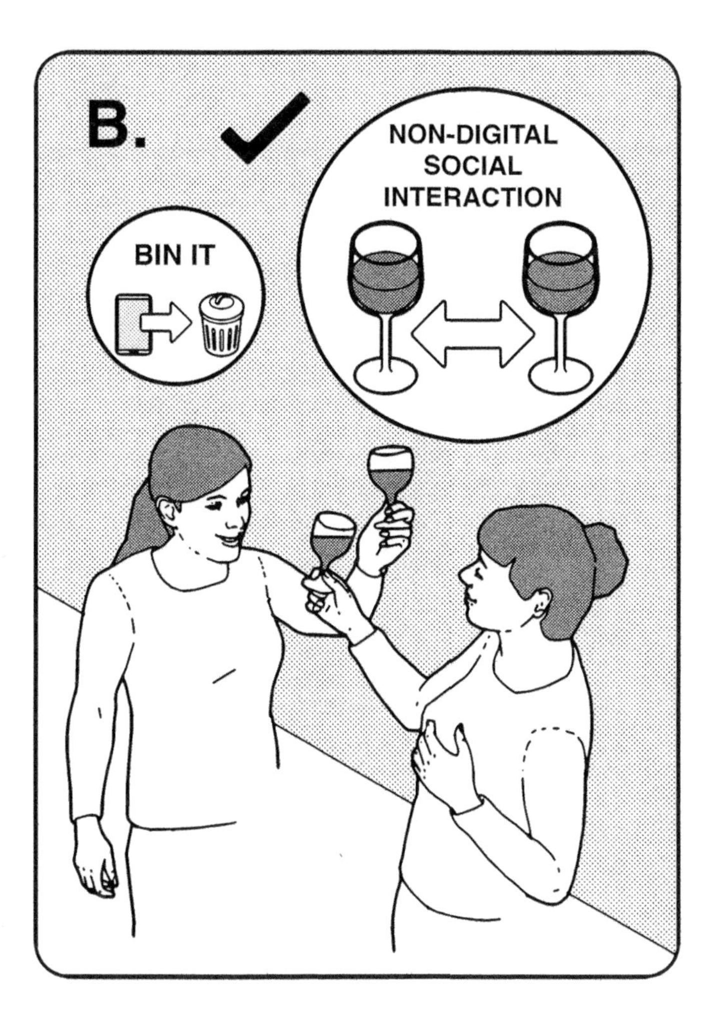
B.
NON-DIGITAL SOCIAL INTERACTION
BIN IT

PARENTING

A.
PERFECT
PARENT.COM
HOTTEST
ACCESSORIES

B.
ENJOY
COMPANY OF
OFFSPRING

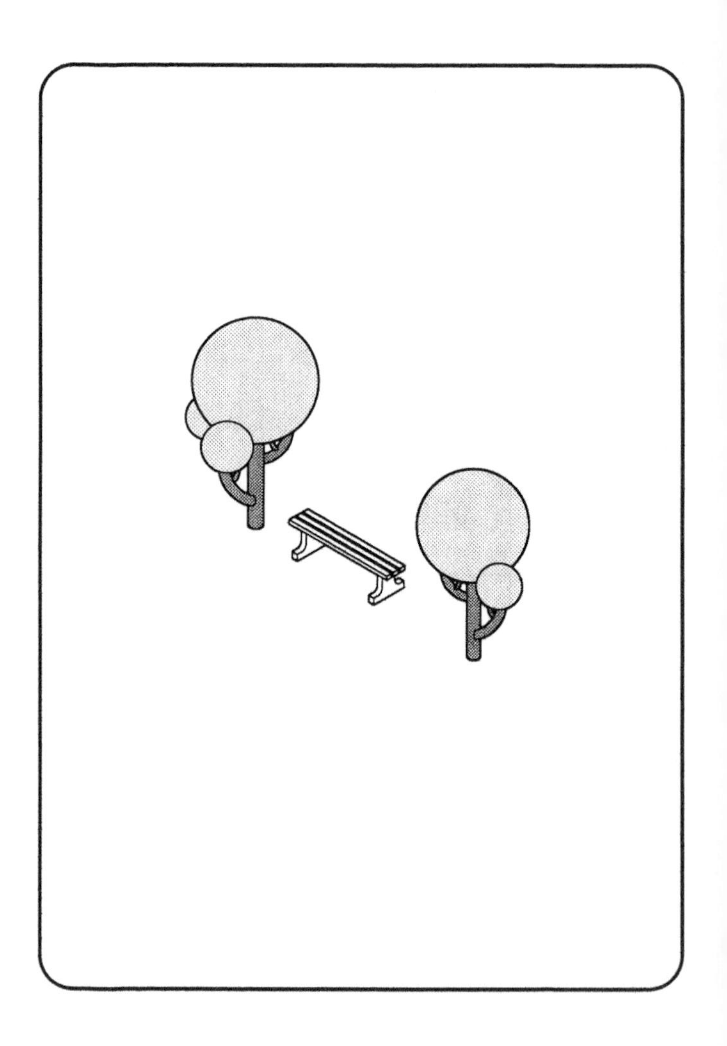

A WALK
IN THE PARK

A.
ZOMBIE
SHUFFLE

B.
BRISK STROLL
ENJOYING PARK

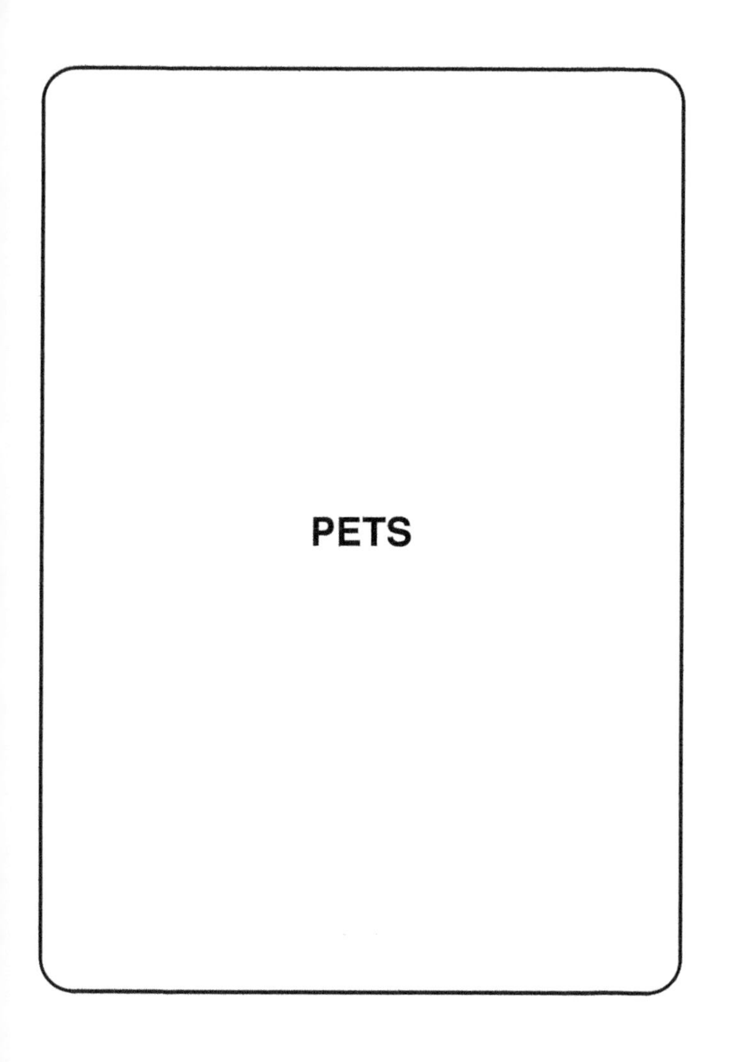

PETS

A.
TRIP HAZARD

B.
LOOK UP
INTERACT

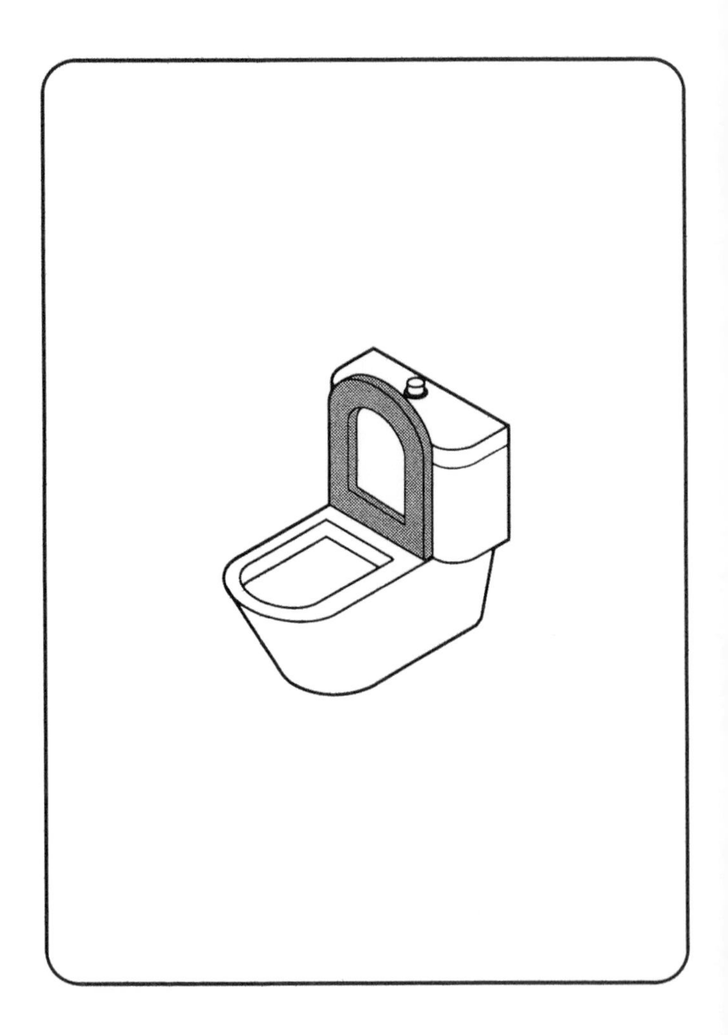

THE BATHROOM

A.
DURATION:
9:32
WIKI:
HOW
TOILET
PAPER IS
MADE

B.
DURATION:
2:28
STARE
AT DOOR

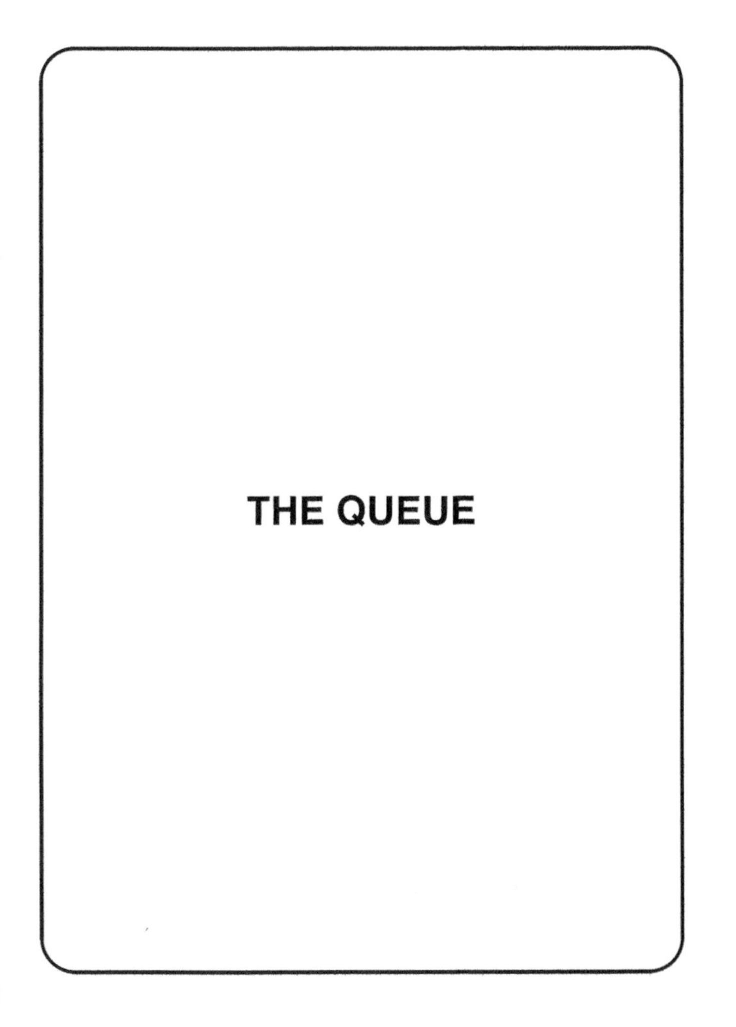

THE QUEUE

A.
HUMAN-LIKE
OBJECT

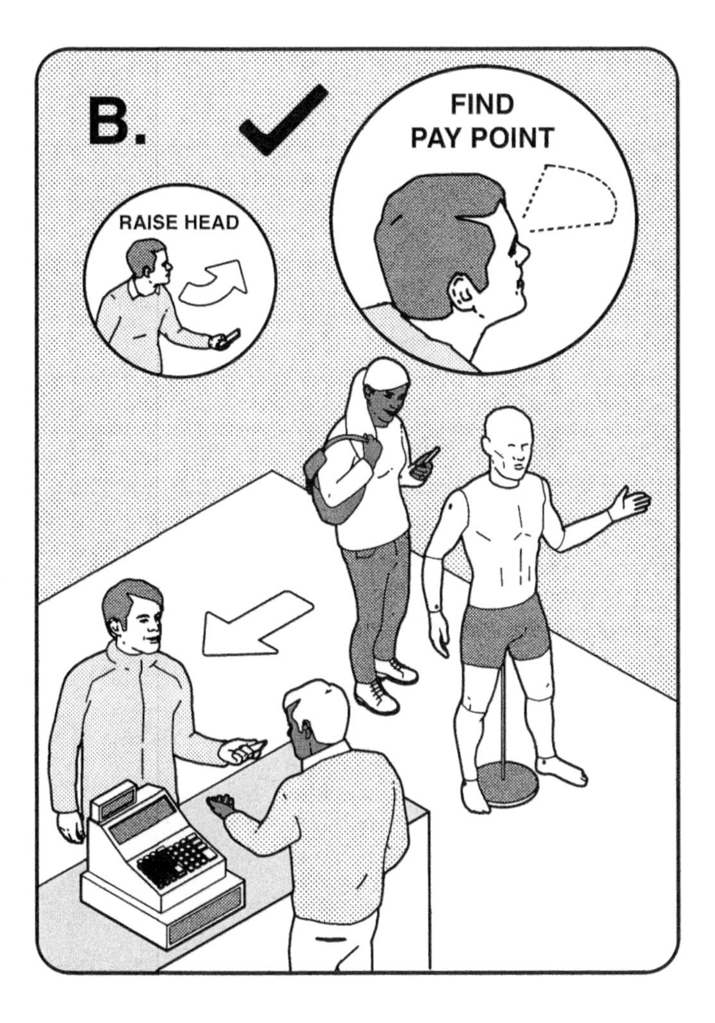
B.
FIND
PAY POINT
RAISE HEAD

ROAD CROSS

A.
SOCIAL
MEDIA CHECK
x15
DANGER
OF DEATH

B.
USE
SENSES

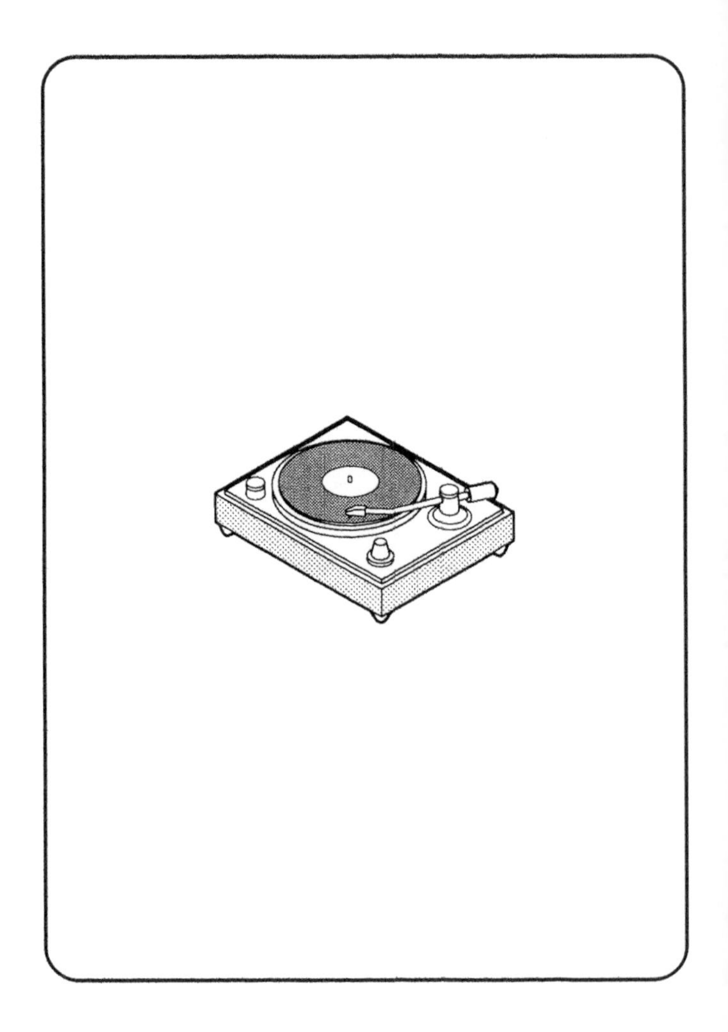

ROCK OUT!

A.
RAINY DAY
PARTY
PLAYLIST

B.
UTILIZE
OLD TECHNOLOGY

PISA
GUIDE

SIGHT SEEING

A.
3 HOURS
TO FULL
CHARGE
PISA
GUIDE

B.
PROCEED
WITH CULTURAL
ENRICHMENT

04:00

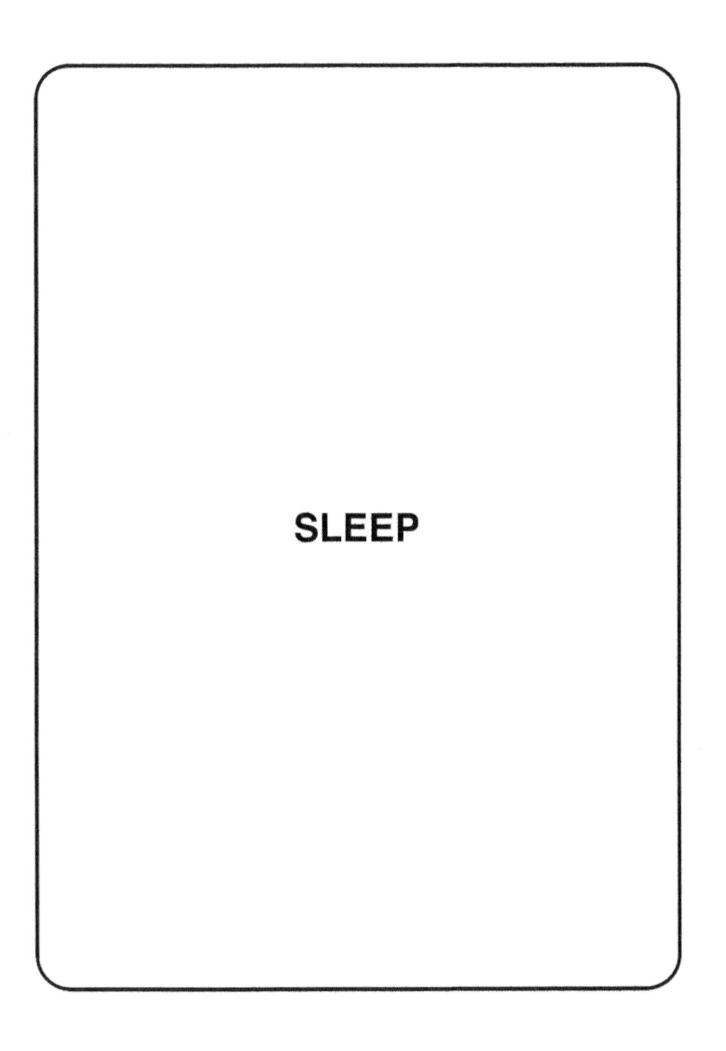

SLEEP

A.
04:00
URGENT!
URGENT!

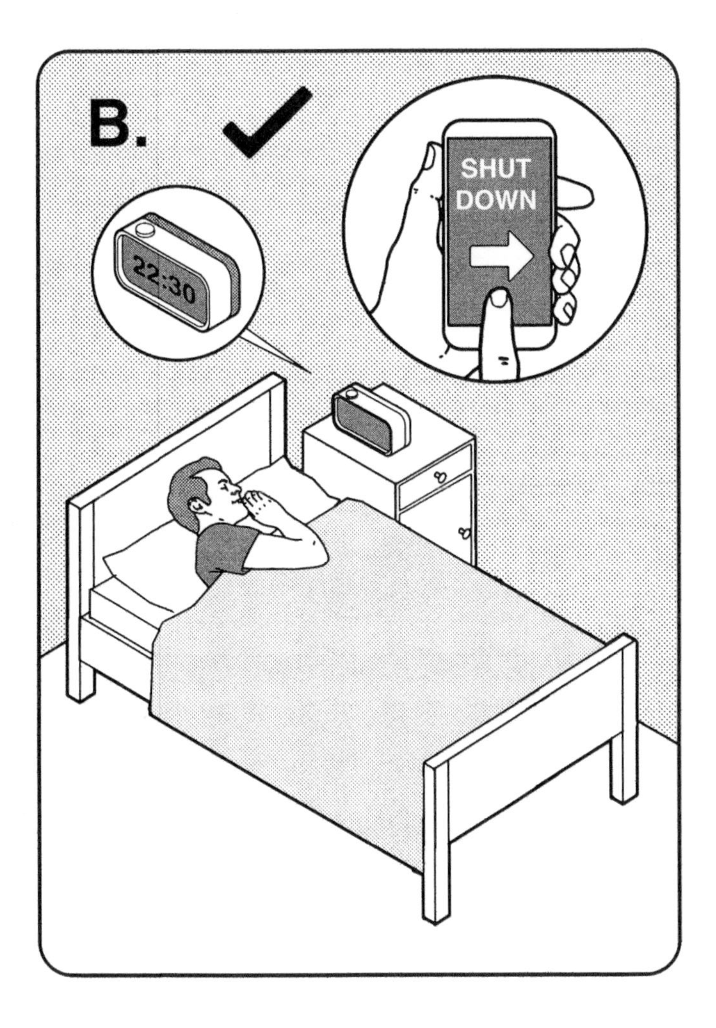
B.
22:30
SHUT
DOWN

THE
STREET HAZARD

A.

B.
OBSERVE
HAZARD
POCKET

THE SUNSET

A.
TWEET:
OMG
#SUNSET

B.
ENJOY THE
BEAUTY OF NATURE

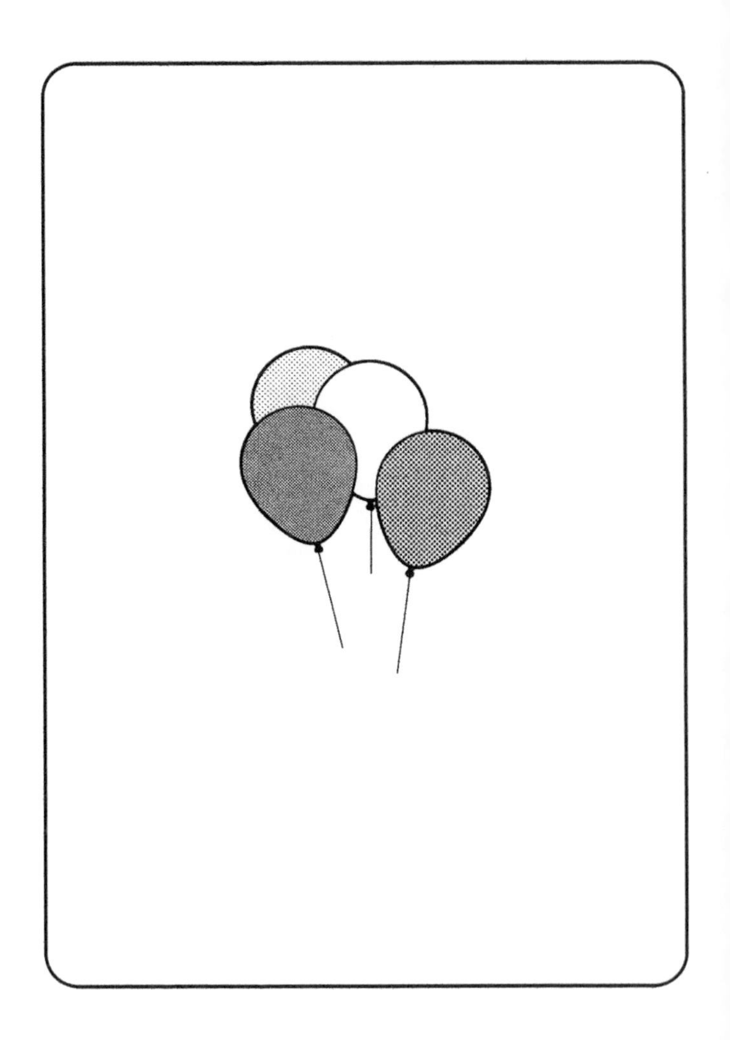

THE SURPRISE

A.
NO
BIRTHDAY
MESSAGES
HAPPY
BIRTHDAY

B.
LOOK UP!
HAPPY
BIRTHDAY

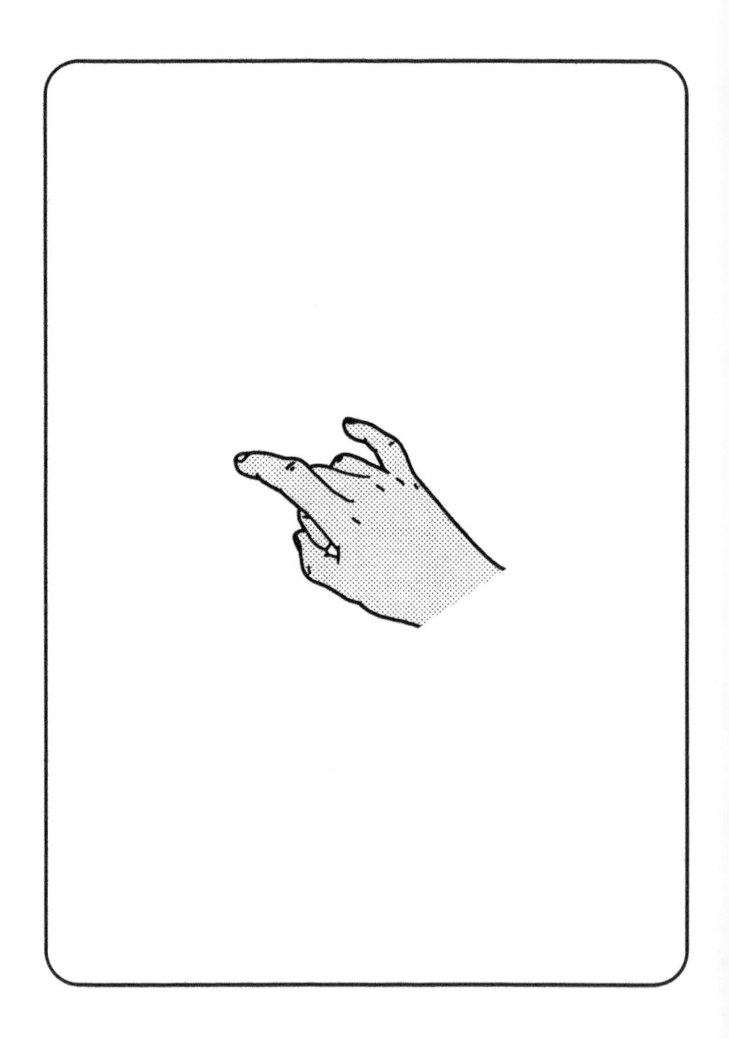

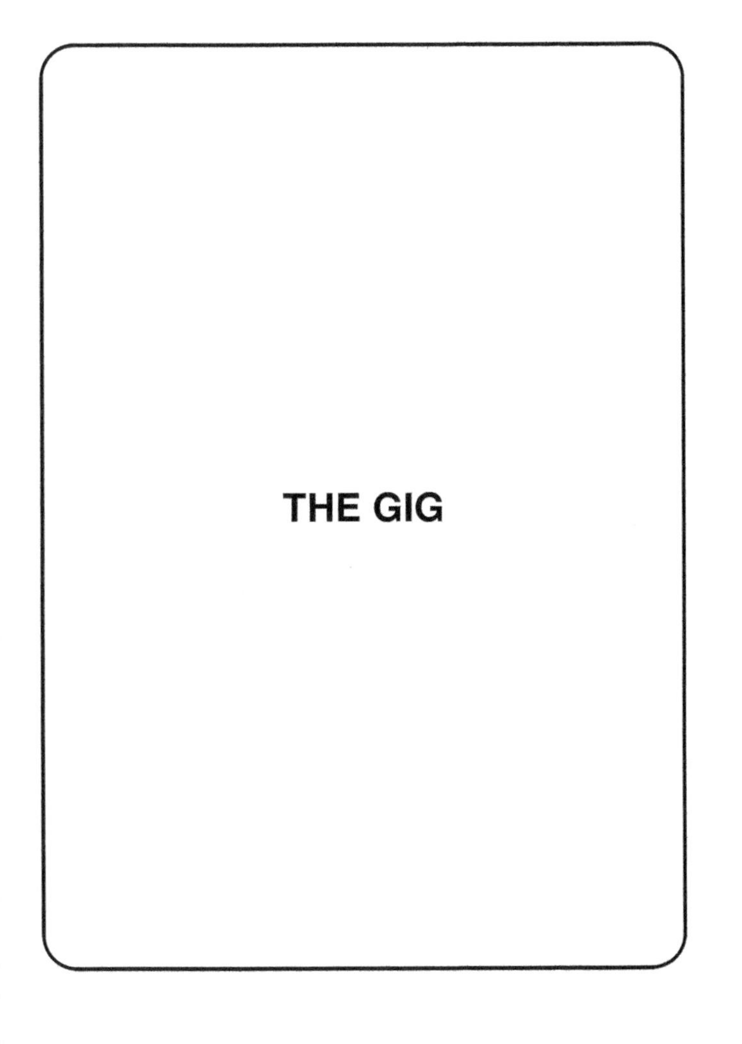

THE GIG

A.
HOLD
STEADY
WHILE
RECORDING
REC

B.
LIVE IN
THE MOMENT
ROCK OUT!

THE LAST TRAIN

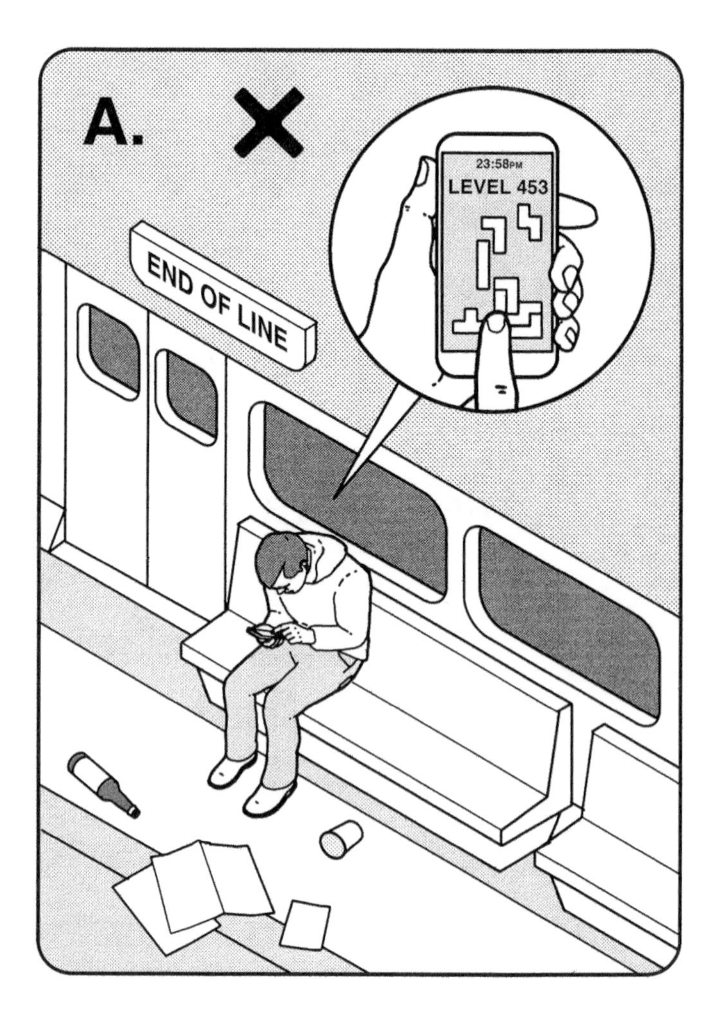
A.
23:58PM
LEVEL 453
END OF LINE

B.
ALIGHT AT
CORRECT STOP
POCKET

THE WATERFALL

A.
SELFIE
OPPORTUNITY

B.
OBSERVE SURROUNDINGS
POCKET

THE WEAK SIGNAL

A.
WEAK
SIGNAL

B.
OBSERVE SURROUNDINGS
LESS ANGER MORE NATURE

THE WEATHER

A.
AM
25°C

B.
USE
“WINDOW”

THE WEDDING

A.
NO
GPS

B.
DO
NOT PANIC
LOOK UP
BIN IT
GET
HITCHED
JUST
MARRIED

SHUT
DOWN